Edmund Goldsmid, Johann Heinrich Heucher

Magic Plants

A Translation of a curious Tract entitled De vegetalibus magicis

Edmund Goldsmid, Johann Heinrich Heucher

Magic Plants
A Translation of a curious Tract entitled De vegetalibus magicis

ISBN/EAN: 9783337372774

Printed in Europe, USA, Canada, Australia, Japan

Cover: Foto ©berggeist007 / pixelio.de

More available books at **www.hansebooks.com**

Magic Plants;

BEING A TRANSLATION OF A CURIOUS TRACT ENTITLED

De Vegetalibus Magicis.

WRITTEN BY

M. J. H. HEUCHER.

(1700.)

EDITED BY

EDMUND GOLDSMID, F.R.H.S.

F.S.A. (Scot.)

PRIVATELY PRINTED   EDINBURGH

INTRODUCTION.

THE curious tract translated in the following pages is one of those strange compositions written in the 17th century, wherein the writer, evidently a man of great erudition, endeavours to unmask what he considers the superstition of his age, but, in his attempt, at once exposes his own credulity.

The belief in Magic and Witchcraft was almost universal in the Middle Ages, and " affected every class of Society, and all sorts and conditions of men." In our own country, as evidenced by our judicial records, the unreasoning credulity which swept across the civilised world, rose to high-water mark, assisted as it was by the efforts of that wise fool of a King, James the First of England and Sixth of Scotland.

Heucher's pamphlet itself is a small 4to. of 16 pp. I know of no copy besides my own, which formed part of the Maidment Library. The imprint is "Lipsiæ, apud Paulum Ehrentium, MDCC."

I have given, as an Appendix, a translation from the Official records of the Royal Court of Guernsey, of the trial of three women for witchcraft in 1617. This translation is taken from Mr. J. L. Pitt's *Witchcraft and Devil-Lore in the Channel Islands.* Any one interested in the subject should get a copy of this very curious, interesting and really well edited pamphlet.

EDMUND GOLDSMID.

EDINBURGH, *March 28th*, 1886.

Magic Plants,

J. M. H. HEUCHER.

1. OFTEN and forcibly does it occur to me to
wonder how it has happened that many men,
of glorious reputation and famous for the greatness
of their illustrious deeds, have believed in wicked
and unlawful arts, yea, in some hidden compact
entered upon with the devil. For, passing over
the dreadful blasphemies of the Hebrew nation
against the Saviour, and the insolence of Hierocles
accusing of sorcery both Him and his most holy
disciples, chiefly Peter and Paul, and also the
impious attacks of Plinius and Apuleius against
Moses, and the false ideas of Theophilus of
Alexandria and Eustathius of Antioch about
Origen, and of the Arians about Athanasius; I
find the great names of Pythagoras, Platinus,
Jamblicus, Porphyrius, Chicus Æsculanus, Anselm

of Parma, Peter of Aponum, Albertus Magnus,
Frischlinus, Henricus Cornelius Agrippa, Michael
Scott, and Roger Bacon, and finally the names of
even the Popes Sylvester II. and Gregory VII.
mentioned in the catalogues of renowned Magi-
cians. And concerning all of these, it is clearly
confirmed that they stood out from the crowd as
illustrious and of great celebrity, as men whom a
lofty spirit and intellect separated from the vulgar
herd of *literati ;* a careful study of astrology and
nature gave them an understanding of all the
most abstruse questions, and transmitted their
brilliant theories even to posterity, so that it
seems to be almost superfluous for any one, after
the example of Gabriel Naudæus, to defend them
by a written apology. But I ceased to wonder
when I thought of envy, the assiduous companion
of virtue, and when I noticed how much it pre-
vails, nay, even when I considered the grounds
that may have led to such belief. There are such
very wicked examples of profane deceivers meeting
us in every age, that I need not go into ancient
history, sacred or profane ; the more recent
writings of Doctor Faust, Zedechias the Jew,
Scotus Parvus, Christopherus Wageners, the three
Eschelles, of Merlinus and Maugis, or of Naudæus
(if you prefer magic romances) suffice. Their
tricks and manifold deceits, although not to be
compared with the wonderful works of wisdom

and art carried out by the learned men mentioned
before, made a deep impression, however, owing
to the particularly gross ignorance of the ages in
which they lived, but they also left on them an
indelible blot of infamy.

2. For such barbarity and want of knowledge
existed in their time, that every class of learning
which was not of the common kind was looked upon
with suspicion, and men, carried away by their own
base ideas, erroneously thought it had been drawn
from the most filthy lake of hell. And, indeed, a
more just or more accurate judgment could scarcely
be expected from those men, since, besides their
utter ignorance of literature, they were also
ignorant of real magic, which embraces in its
sphere the highest culminating point of all sciences,
and by the help of which rare, abstruse, and
unaccustomed effects are produced. For, be it far
from me to assent to the opinion of those who,
with Thomas Erastus and his defender Georgius
Hornius, renounce all magic and condemn it as
illicit and low ; since all those reasons which they
have hitherto brought before the public are open
to many objections, and if they do prove anything,
it is nothing else than that the legitimate use of
magic, as very often happens in other cases, has
been brought into disrepute by abuse. People in
general are hardly cautious enough, and are a little
too violent and abusive regarding lawful magic,

looking upon it as being unlawful and being accustomed to call it forbidden, devilish, deceitful, supernatural, poisonous, black, foul, base, of bad repute, the evil art and other names. These men indeed, believe as Democritus, Averhoes, and Simplicius do, that there are no devils. They either attribute all results to good angels, whom by reason of their superiority they call spirits, and argue that our air is full of them; to which class belong the Platonics, Jamblicus, Porphyrius, Proclus and Julian the Emperor; or else they think, with Avicenna and Petrus Pomponatius his pupil, that all things spring from hidden magnetic properties, and a peculiar state of the atmosphere, or with Reginald Scott trace back most cases to madness, black bile, or other disturbances of nature, or along with Thomas Campanella, and Marcus Marx, believe in the power of concentrated will. I may mention Cornelius Looseus, Gallidus Gaudanus, Robertus Fludd, John Weir, Bodinus, and many others as supporting my contention, since it can be seen by their writings that they believed in such a thing as beneficent sorcery.

3. Therefore, both these opinions being considered, it may be as well to conclude that magic is not altogether to be renounced and despised, since some magic may be found, which is not only free from wickedness, but which even deserves much

commendation. Although I would willingly grant that what is base and devilish, is often blended with that blameless and harmless science, just as that science in an impure state often advances in the guise of one which is legitimate, and adopts methods from nature. This the most ancient of the Magi had already done, as Lucanus, Statius, Seneca the tragedian, and others assert. For they wished to appear to have accomplished every thing they did by the power of Nature. Whence you often read that they invoked Hecate, since by her they specified nature herself, or the Empress of Nature, which fact the passage in Eusebius, in *Præpar. Evang.* makes tolerably clear, wherein Hecate says of herself: "Bearing the threefold symbols of a threefold nature." Nay, what hinders me from stating with Vossius *de Orig. et Pror. Idol.*, that evil spirits accomplish nothing save by natural powers, and that nothing can be done except by the powers of nature. For what, pray, if the power they have of moving themselves and other bodies separate from them, of driving forward or backward the elements, of transporting weights as if by machinery, of gathering things scattered and scattering things gathered, of hiding things seen, of assuming other bodies, of applying actives to passives, of changing organism itself, of destroying sensations, and of accomplishing innumerable other things; what is all this whereby

B

they strive to deceive men, but making use of
of natural forces?

4. All these things have such a considerable bearing on our theme, that they are altogether worthy
of being duly inquired into. There are many
effects which we see produced, particularly
admirable and unusual, which must be attributed
to the Supreme Deity alone, and on that account
you may consistently call them supernatural, since
by a long distance they outstrip the usual powers
of nature. Others indeed, springing from the
same Creator, owe nevertheless their whole
beginning to the motion, imparted and definitely
implanted by fixed laws in the system of the world
at its creation, and are therefore simply called
natural. Others indeed, differ from these to this
extent, that they depend chiefly on our imagination, the stupendous power of which let us not
treat too lightly. And we believe there are some
allied to these, which emanate from superstition,
as Theophastus very well shows in his *Characters*.
Finally, certain results which we are able to assign
to none of these classes of causes yet enumerated, are
attributed to the devil, and thus belong to the
supernatural. Being so far removed from God,
they have been distinguished by the name of
devilish. Moreover, to this list I would add all
those, which although they depend on, and
proceed solely from, some evident natural cause,

nevertheless, because the devil mixes himself up with them, pretends that they are his own work and feigns that he is their author, deserve to be called devilish as well as magical.

5. For, indeed, what powerful mind is not able to perceive by fair investigation, how feebly, how confusedly, and with what small degree of accuracy, these subjects are handled in the writings of most men, for, although we have a very bountiful supply of authors on this theme, so much so that, if you wished to estimate the number of those who have written on magic, your paper would hardly contain them, you will nevertheless be unable to name one who has unfolded and explained everything, as was befitting. I will not, at this time, discuss all magic, which indeed, these little pages would not contain, but I place before myself, as my theme, for the present at least, the vegetable kingdom alone, in which, the industry, if not of all, at any rate of many has failed to enlighten us, and since, so far as I know, former writers have recorded in their writings far too little about magical plants, I will devote myself to this, if I may use the expression, fruitful subject. Poets (since frequent mention of the subject is made in their writings,) must have been aware that thereafter, its usefulness would be fully appreciated. For, who can forget the powerful and noxious herbs of Medea, the enchanting herbs

of Lucan, the fatal herbs of Claudian, the flourish-
ing plants of Maro, and lastly those venomous
plants, of which abundance was to be found in
Colchis and Thessally?

Who is ignorant of the Nepenthe of Homer, the
Polion of Hesiod, the plant of the Ethiopians
drying up waters, the Asphodel, sweet food of the
Manes, the plant of Juba and Bela arousing the
dead, the soothsaying Thelangis, the Juniper and
Hypericum putting the demons to flight, the
Alyssus, Baccharis, Adianthus, Satyrion and the
Amulets of the Syderii charms against witchcraft? Nevertheless it has not yet appeared
necessary to any one to explain these mysteries so
that the observant reader might rest satisfied with
regard to them. Which fact therefore the more
impels me to embrace the opportunity.

6. I confess that, at first, I was filled with no
inconsiderable joy when I read the list of the
authors who have spoken of magic, not without
praise—Hermes, Orpheus, Musæus, Pythagoras,
Democritus, Empedocles, Kiranus, and others, all
of whom are said to have praised magical plants ;
from this I promised myself a great solace to my
labour. But, indeed, all hope immediately came
to naught when I found that most of these authors
had been destroyed by the injury of time, or if
any fragments of their writings survive they are of
uncertain and slight reliance, as there is still

much doubt as to the real authorship of such fragments. Nor do the writers whose works on plants survive satisfy my expectations much more, —namely, Theophrastus Eresius, Plinius, Apuleius, and those who have drawn small streams from these sources—Theophrastus, Paracelsus, Th. Cantipratanus, the feigned name of Albert the Great, Bernh. G. Penotus, Alex. Suchtenius, Leonh. Turnheuserus, Carrichterus, Mich. Sendigovius, Jo. Ern. Burggravius, Wreckerus, Porta, Lemnius, and a few others. These writers on plants, on the wonders of nature and other hidden things, however much they may wish to appear to have founded the healing effects of herbs on natural laws, either wittingly or corrupted by superstition, have nevertheless inserted in their writings certain things which savour somewhat of magic and supernatural agency.

7. It behoves the mind to approach this subject free from every preconceived notion and superstition, and properly imbued with natural science and learning, nay, even with a knowlege of the healing art itself. For when those who first wrote on the subject gave their attention almost solely to meditation about the stars, and believed natural philosophy and the healing art to be intimately connected with these heavenly bodies, they disseminated seeds of magic, from which that science grew to such a degree that it over-

spread the whole world with its pollution. Ot
this fact Pliny is a remarkable witness in his
Natural History, confirming what we have said
by these words :—" No one can doubt that magic
is the greatest of the sciences, seeing that it is the
only one which embraces three other sciences
having power over the human mind, and reduces
them to one. No one doubts that it has sprung
from medicine, and become something loftier and
holier than its parent." This is what Grillus also
maintains, saying that all the Egyptians are
physicians. From which fact it is clearly proved
that the healing art almost always flourished
among that nation, and thus, as we shall soon
show, contributed much to our argument.

8. I will not flatter myself that this work will be
so complete as to appear perfect in all parts before
the learned world. It is sufficient to have wished
to make it so, in so far as it goes. I have written
this present work on the so-called vegetable
kingdom, including and embracing in its limits
nothing less than all that springs from the soil,
and is thus brought into existence, to germinate,
be nourished and increase, both in roots, leaves,
stalks, stems, flowers, branches, twigs, fruits and
seeds; in short, nothing in this work belongs
either to the animal or mineral kingdoms, and let
it be remembered that magic specifies not merely
what is devilish, but also all that which is con-

sidered supernatural through ignorance and superstition. Since therefore, I intend to treat of
magical herbs, I will specially enumerate if not
all, at least most plants, herbs, trees, leaves,
flowers, fruits and seeds, which are considered to
savour of anything superstitious and deceitful, and
those by means of which the great author of
falsehoods imposes on men ; and then, as far as I
can, I will enquire into the causes of the effects
attributed to these herbs.

And, from this little work of mine, I hope not
only that some advantage will accrue to those who
long to understand many of the more obscure
passages of old authors, but I also believe that
many of the deceptions practised by sorcerers will
be detected, by which it is possible that our
knowledge of natural philosophy, as well as of
medicine may greatly be increased.

9. Wherefore, if any one is resolved to commence
from the beginning, he, in my opinion, will not
be far wrong if, first of all, he seek the origin of
science in the study of old authors on the healing
art. Thus, although the first principles of the
science ought rightly and deservedly to be attributed to the Egyptians — whose very ancient
records, however, are far from clear—yet these
elements are by no means free from superstition,
for Africa is considered the mother of all unna-
tural things. Different, indeed, was the reasoning

of the Egyptians from that of Aristotle. For he,
dealing with the interpretation of dreams, decided
that no dreams were sent by the Gods to men, as
it was unworthy of their dignity to go into the
couches of sleepers and disturb the weary. The
Egyptians, on the contrary, when they were about
to consult destiny regarding the cure of any severe
and particularly desperate disease, having duly
purified their mind before they betook themselves
to the couch, used to wait for the revelations of
the Gods in their dreams, and according to the
dictates of these revelations, used this or that
method of cure. As this rite was a little too
simple at first, so in process of time, it was more
and more overloaded with ceremonies. For,
when many plants and flowers had come under
the notice of the sorcerers, they, with the selfish
imagination of enthusiasts grew to consider them
as favours stored up beforehand for themselves.
Then they besought heaven and the Gods with
hymns, to show by means of a dream, which
plants were the best cure for the disease, then,
after preparing themselves by prayer to sleep,
whatever deity first appeared to them in their
dream was supposed to indicate the plant specially
attributed to him, as the one divinely destined to
subdue the disease. Which fact Jamblicus,
referring to the Egyptian mysteries, testifies to in
the following words : Thus, in the temple of Escu-

lapius, diseases are cured by divine dreams, and the healing art arose from holy dreams." And Aristides further confirms this.

10. Not that the Egyptians, at the shrines of Isis, Osiris, and of the Egyptian Aesculapius, which was close by the banks of the Nile, followed this custom in the same manner as the Greeks and Romans in the temples of Hephæstion or Vulcan, but the general process was as familiar to the Greeks and Romans as to the Egyptians. For, with this intention and design, the former often used to sleep in the temple of Æsculapius, to which custom Aristophanes refers when he says : "Let both you and I go as soon as possible, and lie down in the temple of Æsculapius." Which rite Plautus also alluded to in Latin ; "Therefore it happens that this sick procurer lies in the temple of Æsculapius." They were accustomed to act in the same way in this temple, as in the city of Pergamos in Asia, whither Herodianus states that Antoninus Caracalla had set out, "wishing to use the cures of Æsculapius." The same thing used to take place in the temple of the Oropian Amphiarus, about which Pausanias says : "Sacrificing a ram to him, and spreading out the skin to sleep, waiting for the appearance of the dream." And from these testimonies of authors it is pretty clear, how remarkably the "lying down" of the Greeks, and the zealously "watching the gods"

c

of the Latins, the hides of those, and the couches
of these, corresponded to one another, all of
which things have reference to the custom of
sleeping in temples and to the rite of lying in wait
for divine revelations to be obtained by means of
sleep. In very truth, what class of vulgar dreams
these were, those may judge, who know how
perverse and idolatrous a worship of the gods
prevailed in those times, and what sort of deities
the heathen worshipped and honoured. I, indeed,
have no doubt that the devil was by no means
forgetful of himself, but was chiefly intent on his
own profit, and that he had embraced this suitable
and very convenient opportunity of defiling the
natural and healing Science by means of the most
wicked superstitions. Which thing even Borrichius in
other passages the keenest opponent of Couringuis,
who thinks along with us, considers necessary to
confess in his book on the hermetic art, and on
account of this, thinks that we should lament the
miseries of these nations estranged from the true
worship of the deity. Nevertheless, Kircherus in
his Œdipus Ægyptiacus, seems to have spoken
too strongly about these enchantments of dreamers,
since he thought that, although this plan of curing
afforded wonderful results even in lamentable
diseases, these were by no means to be ascribed
to the natural power and quality of the plants, but
to a hidden compact and treaty with demons. For

although very often cures have been effected, the plants used in which have been only an outward sign, the devils supplying the actual power of curing, no one, nevertheless, will hesitate to assert that none of those remedies, acting in a natural way, have displayed a power and efficacy of their own.

11. And so the Egyptians obtained some fair knowledge of plants, and did not remain content with dreams, but advanced even further. They had, indeed, selected from those who had appeared to the dreamers ten gods, whom they used especially to entreat to show the readiest cures of diseases. To each of these they had given a separate number — number one to the Supreme Deity and the Cause of Causes, number two to Mephta, number three to Minerva, number four to Apollo or Horus, number five to Isis, number six to Osiris, number seven to Mercury, number eight to Ammon, number nine to Typhon, lastly, number ten to Momphta, and they had thus come to connect numbers with the gods. Thereafter they examined the numbers the leaves, plants, and flowers had, how many pips fruit contained, and how many joints stems possessed, and afterwards considered them sacred to that god whose number was thus represented, and placed them under his guardianship. Nor did they attend to the number only, but earnestly watched the colour

and the form in flowers ; and as regards stalks of
plants, they duly inquired whether they were
round, smooth, three-sided or four-sided, and
dedicated these stalks as sceptres to the corres-
ponding deities, as can be further proved from the
Bembina Fabula in the writings of Kircherus.
They used to believe that they were able to show
wonderful results, which fact the sublime Platonic
Proclus, in his book concerning sacrifices and
magic, besides others, confirms. He adds that
by amalgamation of many things they obtained
supernatural influence, but he thinks, nevertheless,
that sometimes one plant or one stone has been
sufficient for the divine work, adding that in stones
and plants is concentrated the essence of divine
powers.

12. Further, the chemical plants, which (if we
believe Origen and Stobæus) were thirty-six in
number, clearly prove this very thing. For,
dividing the whole human body into so many
parts, they assigned a separate divinity to each,
whether god or devil ; or, with Galenus and J.
M. Firmicus, you may call them decani ; each of
whom was the presiding genius who looked after
the safety of his own appointed limb or member.
And if a man lost the health of any member they
persuaded themselves that it could be recovered,
its respective deity being invoked by the name
peculiar to himself, and propitiated by the plant

dedicated to himself. Firmicus, Hæphestion the Theban, and Manlius differ among themselves about the Decani and their number, whether there were thirty-six only, three always being given to each sign of the zodiac, or whether, indeed, there were more ; or whether every one of these had under him his own subordinates and fellow-workers ; I am content to believe that they were divided according to the twelve signs of the zodiac, possessed of unlimited power over the whole human body, and were wont to be propitiated by certain plants. Nay, it was a proper and extremely common custom to appoint to separate signs—nay, to separate constellations—their own herbs ; Abenrahman, the Arabian philosopher, moreover, relates that they chose for use as medicines all forms of nature and grades of beings which possessed any resemblance to, or quality in common with, the seven planets. As an example, all forms which bore any resemblance to the sun they chose from minerals, metals, stones, animals, plants, and herbs, and turned them to use as medicine for the heart, because the heart is, as it were, the sun of a smaller world ; which things can be still more comprehensively read in Jamblichus's *De Mysteriis Ægyptiorum*, Marsilius, and likewise in Ficinus.

13. Who does not see that later, further superstitions sprang from this connection of plants with Heavenly

bodies, and amongst these I may specially mention
the belief that the efficacy of the plants depended
on the hour they were gathered in the meadows,
which hour was fixed by the influence of the stars.
And those will not oppose this base and false
custom, who still believe that greater powers lie
in the stems of Hypericum, dug up on the Feast
of St. John, in twigs of the wild cherry cut off on
the sacred anniversary of St. Martin, and in the
sympathetic ashen wood, split up at noon, when
the sun is in the sign Virgo, and the moon in a
crescent state.   Of this superstition Keinsius
(in var. lect. Freitag. noct. med.), Maxwellus
(Medicin. Magnet.) and Grabius *(Arcan. medic.)*
supply many examples.   Kircherus tries to prove
that, particularly at the time of the solstice, the
leaves of the willow, white poplar, elm, lime, and
olive turn their under side to the sky; for this very
reason, therefore, I am not willing to leave out of
account all motions of the heavenly bodies;
besides, in the work of Gesnerus, we come across
not a few examples of plants having been affected
in shape by the sun and moon, this being a clear
proof of harmony with their own constellations.
Much less do I strive to reject or reproach all
plucking of plants at fixed times as being an
unnecessary process, desiring merely to prove, with
Couringius, that the idea of the risings, settings,
and positions of the constellations fixing the

different times of collection, is ridiculous, and
moreover, that the belief that we can determine
the very days, hours, and even minutes, savours
altogether of superstition, and by no means differs
from the doctrines of the Egyptians. Which
ideas Helmontius refutes uncommonly well, and
censures in the following words. He says, "In
modern books on Medicine there be some who
ascribes a great number of diseases to signs of the
Zodiac, and since the number of these signs was
too small, they extended each into three sections,
in order, forsooth, to divide all herbs into thirty
six classes. Do we not know that the soil has a
capacity of raising plants from itself, which capacity, therefore, it need not beg from heaven."
Finally, concerning the apportionment of plants to
the seven planets I have nothing to say, except
that the very number of the planets now known
sufficiently silences the theory.

14. Neither, however, is it right, in making
researches about plants, to believe that all the
science of the Egyptians was magic; although I
cannot but greatly approve of the way they judged
of the excellence of herbs by the help of their
external senses. Nay, even he may not have
erred who has said that these men surpassed in
many ways the carefulness of later writers, when,
besides taste and smell, which almost alone you
see our botanists employ, they also summoned to

their assistance hearing, sight, and touch. With
the help of the ear they judged by the sharp,
heavy, ringing, dull, harsh, or gentle sound, of
the consistency of the stem ; by the hand, from
the smoothness, roughness, softness, or hardness,
of its texture; and lastly, by the eyes they decided
from the colours as to the virtues lurking within.
Thus, by judging carefully from their senses, they
were able to infer and learn the effects of plants.

15. But, truly, the more correctly they conducted
themselves in this matter the more basely did
they devote themselves to superstition. Carefully
they used to watch for the distinctive mark of
every plant, observed to which member of the
human body it corresponded, and carefully noted
the result, afterwards thinking that plant most
suitable to that limb which it resembled by its
own external shape. Or by a certain supposed
analogy of qualities, they applied it as a remedy
to that limb. Thus the peony flowers yet enclosed
in the bud and poppy heads they dedicated to the
head. In like manner the Euphrastia, Caltha,
Heiracius, and Anthemis they thought beneficial
to the eyes, because in them they found something
of the first elements of the eye : they assigned the
root of the Dentaria to the teeth, of the Anthora
to the heart. If, therefore, they came across any
plant, the Cynosorchis, for example, like the
genitals in shape, they persuaded themselves that

it most certainly cured the sexual passions, since by its influence they believed that deity could be summoned who presided over the genitals.

16. For a like reason, also, they employed the juices of plants which in colour resembled the juices or humours of the human body, for the purpose of cleansing it of the offending humour. Hence, they wished to cure the yellow bile with the saffron-coloured juices of plants, the black bile with the black, purple or azure phlegm with the white, blood with the red, and milk and semen with the juices resembling milk. Whence it can be inferred that this science was not, as many (among whom is Couringius) think, invented in modern times, but that it was raised afresh from the records of Eastern people, since very distinct proofs of this are found in the writings of Dioscoris and Pliny, and, probably, from the absurd customs of the Egyptians. I confess that, on this point, I rather prefer now to suspend my judgment than to settle on anything as certain. For even if the opponents of this theory maintain that it follows as a consequence from it that the fruit of the Anacardium, for example, must be the best cordial, that the famous Cephalic must produce the best apples; the Aphrodisiac, the best plums; and the juice of the Esula the best milk; they nevertheless prove nothing from that, as it can be at once retorted: How long is it since Mercury,

D

Antimony, Cinabar, Coriander seeds, &c., were excluded from the list of poisons? Borriclius argues as follows : It is a characteristic of human, not of Divine nature, to inquire remissly into things. Nothing has ever been discovered or created until after many weak and apparently futile efforts.

17. Those are the original fibres, as it were, from which the tree of the belief in the magic virtues of plants grew to such an extent, as to spread its branches so widely that it refuses to be embraced in these narrow limits. It is a matter of enquiry even whether the land of the Egyptians supplied itself with its own magic plants. Couringius shows that it furnished a few wines, but did not abound in medicinal remedies. Let writers contend among themselves as long as they wish, we refer those who desire to know more to the books on Egyptian plants published by Kircher and other writers, from which each can form his own opinion. But now, I regret that I am prevented by the limits of my parchment from entering into any disquisition concerning the merits of these plants, whose names I have recorded. There are indeed among their number some of such a nature, that I may briefly mention their names and properties: the Salvia, Veratrum, Juniperus, &c,, powerful in routing the demons; the Adianthus, Ruta, Sideritis, &c., a sovereign charm for over-

coming witchcraft; the Absinthium, Ricinus, Cnebison, Scordotis, &c., used in sacrifices, and to obtain divine apparitions; the Asphodel, irresistible in calling forth the gloomy deities and the Manes; the Osiris, &c., which enables one to prophesy; the Laurus, Theangelis, Halicaccabi, Bellonasia, Vatica, &c., which bring on madness, ravings and sleep; Strichnon, Thallassegle, Gelotophyllis, Semasum, the plant of Nectanebus, the Olive, Mandragora, Catanance, Cemos, Anacampserotis, &c., which enter into the composition of aphrodisiacs or love-potions. The Heiracium, Theombrotium, Bali, Juba, &c., said to have power to rouse from death itself. I shall with care explain, time and opportunity permitting, the qualities of these and many other plants, but meanwhile I must bid thee farewell, benevolent reader.

Laus Deo.

APPENDIX.

---><<>><---

CONFESSIONS OF WITCHES UNDER TORTURE.

Before AMICE DE CARTERET, Esq., Bailiff,
and the Jurats.

JULY 4th, 1617.

SENTENCE OF DEATH.

Collette du Mont, widow of *Jean Becquet*; *Marie,* her daughter, wife of *Pierre Massy*; and *Isabel Becquet,* wife of *Jean Le Moygne,* being by common rumour and report for a long time past addicted to the damnable art of Witchcraft, and the same being thereupon seized and apprehended by the Officers of His Majesty [James I.], after voluntarily submitting themselves, both upon the general inquest of the country, and after having been several times brought up before the Court, heard, examined, and confronted,

upon a great number of depositions made and produced before the Court by the said Officers; from
which it is clear and evident that for many years past
the aforesaid women have practised the diabolical art
of Witchcraft, by having not only cast their spells
upon inanimate objects, but also by having retained
in langour through strange diseases, many persons and
beasts ; and also cruelly hurt a great number of men,
women, and children, and caused the death of many
animals, as recorded in the informations thereupon
laid, it follows that they are clearly convicted and
proved to be Witches. In expiation of which crime
it has been ordered by the Court that the said women
shall be presently conducted, with halters about their
necks, to the usual place of punishment, and shall
there be fastened by the Executioner to a gallows,
and be hanged, strangled, killed, and burnt, until their
flesh and bones are reduced to ashes, and the ashes
shall be scattered ; and all their goods, chattels, and
estates, if any such exist, shall be forfeited to His
Majesty.   In order to make them disclose their
accomplices, they shall be put to the question before
the Court, previous to being executed.

Sentence of Death having been pronounced against *Collette Du Mont,* widow of *Jean Becquet; Marie,* her daughter, wife of *Pierre Massy;* and *Isabel Becquet,* wife of *Jean Le Moygne;* the same have confessed as follows :—

CONFESSION OF COLLETTE DU MONT.

First, the said *Collette* immediately after the said sentence was pronounced, and before leaving the Court, freely admitted that she was a Witch; at the same time, not wishing to specify the crimes which she had committed, she was taken, along with the others, to the Torture Chamber, and the said question being applied to her, she confessed that she was quite young when the Devil, in the form of a cat :* appeared to her : in the Parish of Torteval : as she was returning from her cattle, it being still daylight, and that he took occasion to lead her astray by inciting her to avenge herself on one of her neighbours, with whom she was then at enmity, on account of some damage which she had suffered through the cattle of the latter ; that since then when she had a quarrel with anyone, he appeared to her in the aforesaid form : and sometimes

---

* As regards these colons, occurring where they are not required, Mr Pitts observes that they correspond to similar pauses in the original records, and evidently indicate the successive stages by which the story was wrung from the wretched victims. They are thus endowed with a sad and ghastly significance, for it must be remembered that the confessions were not made in a connected form, but were elicited by leading questions, often accompanied by a fresh spell of torture.

in the form of a dog : inducing her to take vengeance upon those who had angered her : persuading her to cause the death of persons and cattle,

That the Devil having come to fetch her that she might go to the Sabbath, called for her without any-one perceiving it : and gave her a certain black ointment with which (after having stripped herself), she rubbed her back, belly and stomach : and then having again put on her clothes, she went out of her door, when she was immediately carried through the air at a great speed : and she found herself in an instant at the place of the Sabbath, which was sometimes near the parochial burial-ground : and at other times near the seashore in the neighbourhood of Rocquaine Castle : where, upon arrival, she met often fifteen or sixteen Wizards and Witches with the Devils who were there in the form of dogs, cats, and hares : which Wizards and Witches she was unable to recognise, because they were all blackened and disfigured : it was true, however, that she had heard the Devil summon them by their names, and she remembered among others those of *Fallaise* and *Hardie ;* confessed that on entering the Sabbath : the Devil wishing to summon them commenced with her sometimes. Admitted that her daughter *Marie,* wife of *Massy,* now condemned for a similar crime, was a Witch : and that she took her twice to the Sabbath with her : at the Sabbath, after having worshipped the Devil, who used to stand up on his hind legs, they had connection with him under the form of a dog ; then they danced back to back. And after having danced, they drank wine (she did not know what colour it was), which the Devil poured out of a jug into a silver or pewter goblet ; which wine did not seem to her so good as that which was usually drunk ; they also ate white bread which he presented to them—she had never seen any salt at the Sabbath.

Confessed that the Devil had charged her to call, as she passed, for *Isabel le Moygne* : when she came to the Sabbath, which she had done several times. On leaving the Sabbath the Devil incited her to commit various evil deeds : and to that effect he gave her certain black powders, which he ordered her to throw upon such persons and cattle as she wished ; with this powder she perpetrated several wicked acts which she did not remember: among others she threw some upon *Mr Dolbell*, parish minister : and was the occasion of his death by these means. With this same powder she bewitched the wife of *Jean Maugues:* but denied that the woman's death was caused by it : she also touched on the side, and threw some of this powder over the deceased wife of *Mr Perchard*, the minister who succeeded the said *Dolbell* in the parish, she being *enceinte* at the time, and so caused the death of her and her infant—she did not know that the deceased woman had given her any cause for doing so.

Upon the refusal of the wife of *Collas Tottevin* to give her some milk : she caused her cow to dry up,

E

by throwing upon it some of this powder : which cow she afterwards cured again by making it eat some bran, and some terrestrial herb that the Devil gave her.

CONFESSION OF MARIE BECQUET.

Marie, wife of *Pierre Massy*, after sentence of death had been pronounced against her, having been put to the question, confessed that she was a Witch; and that at the persuasion of the Devil, who appeared to her in the form of a dog : she gave herself to him : that when she gave herself to him he took her by the hand with his paw : that she used to anoint herself with the same ointment as her mother used : and had been to the Sabbath upon the bank near Rocquaine Castle with her, where there was no one but the Devil and her as it seemed : in the aforesaid form in which she had seen him several times : She was also at the Sabbath on one occasion among others in the road near *Collas Tottevin's ;* every time that she went to the Sabbath, the Devil came to her, and it seemed as though he transformed her into a female dog ; she said that upon the shore, near the said Rocquaine : the Devil, in the form of a dog, having had connection with her, gave her bread and wine, which she ate and drank.

The Devil gave her certain powders: which powders he put into her hand, for her to throw upon those whom he ordered her : she threw some of them by his orders upon persons and cattle : notably upon the child of *Pierre Brehaut.* Item, upon the wife of *Jean Bourgaize,* while she was *enceinte.* Item, upon the child of *Leonard le Messurier.*

CONFESSION OF ISABEL BECQUET.

Isabel, wife of *Jean le Moygne,* having been put to the question, at once confessed that she was a Witch: and that upon her getting into a quarrel with the woman *Girarde,* who was her sister-in-law: the Devil, in the form of a hare, took occasion to tempt her: appearing to her in broad daylight in a road near her house: and persuading and inciting her to give herself to him: and that he would help her to avenge herself on the said *Girarde,* and everybody else: to which persuasion she would not at the moment condescend to yield: so he at once disappeared: but very soon he came again to her in the same road, and pursuing his previous argument: exhorted her in the same terms as above: that done, he left her and went away, after having previously given her a sackful of parsnips; she then took a certain black powder wrapped in a cloth in which he placed it; which powder she kept by her. He appeared to her another time under the same form in the town district, inciting her anew to give herself to him, but she not wishing to comply, he next made a request to her to give him some living animal: whereupon she returned to her dwelling and fetched a chicken which she carried to him to the same place where she had left him, and he took it: and after having thanked her he made an appointment for her to be present the next morning before daylight at the Sabbath, promising that he would send for her: according to which promise, during the ensuing night, the old woman *Collette du Mont,* came to fetch her, and gave her some black ointment, which she had had from the Devil; with this (after having stripped herself) she annointed her back and belly, then having dressed herself again she went out of her house door : when she was instantly caught up : and carried across hedges and bushes to the bank on the sea shore, in the neighbourhood of Rocquaine Castle, the usual place where the Devil kept his Sabbath ; no sooner had she arrived there than the Devil came to her in the form of a dog, with two great horns sticking up : and with one of his paws (which seemed to her like hands) took her by the hand : and calling her by her name told her that she was welcome : then immediately the Devil made her kneel down : while he himself stood up on his hind legs ; he then made her express detestation of the Eternal in these words : *I renounce God the Father, God the Son, and God the Holy Ghost ;* and then caused her to worship and invoke himself in these terms : *Our Great Master, help us !* with a special compact to be faithful to him ; and when this was done he had connection with her in the aforesaid form of a dog, but a little larger : then she and the others danced with him back to back : after having danced, the Devil poured out of a jug some black wine, which he presented to them in a wooden bowl, from which she drank, but it did not seem to her so good as the wine which is usually drunk : there was also bread— but she did not eat any : confessed that she gave herself to him for a month : they returned from the Sabbath in the same manner that they went there.

The second time she was at the Sabbath was after the old woman *Collette* had been to fetch her, and she anointed herself with the ointment as above stated ;— declared, that on entering the Sabbath, she again had connection with the Devil and danced with him; after having danced, and upon his solicitation to prolong the time, she gave herself to him for three years ; at the Sabbath the Devil used to summon the Wizards and Witches in regular order (she remembered very well[1] having heard him call the old woman *Collette* the first, in these terms : *Madame the Old Woman Becquette):* then the woman *Fallaise;* and afterwards the woman *Hardie.* Item, he also called *Marie,* wife of *Massy,* and daughter of the said *Collette.* Said that after them she herself was called by the Devil : in these terms : *The Little Becquette:* she also heard him call there *Collas Becquet,* son of the said old woman (who [*Collas*] held her by the hand in dancing, and someone [a woman] whom she did not know, held her by the other hand): there were about six others there she did not know : the said old woman was always nearest to the Devil : occasionally while some were dancing, others were having connection with the

Devils in the form of dogs; they remained at the Sabbath about three or four hours, not more.

While at the Sabbath the Devil marked her at the upper part of the thigh: which mark having been examined by the midwives, they reported that they had stuck a small pin deeply into it, and that she had not felt it, and that no blood had issued: she did not know in what part the Devil had marked the others: those who came first to the place of the Sabbath, waited for the others; and all the Wizards and Witches appeared in their proper forms: but blackened and disfigured so that they could not be recognised.

The Devil appeared sometimes in the form of a goat at the Sabbath; never saw him in other forms: on their departure he made them kiss him behind, and asked them when they would come again: he exhorted them always to be true to him: and to do evil deeds, and to this end he gave them certain black powders, wrapped in a cloth, for them to throw upon those whom they wished to bewitch: on leaving the Sabbath, the Devil went away in one direction and they in the other: after he had taken them all by the hand: At the instigation of the Devil she threw some of the powder over several persons and cattle: notably over *Jean Jehan*, when he came to her house to look for a pig. Item, over the child of *James Gallienne*, and over others. Item, over the cattle of *Brouart*, and of others.

It was the Devil that was seen at the said *Gallienne's*

house in the form of a rat and a weazle, she herself
being then in the neighbourhood of *Gallienne's* house,
and he [the Devil] came to her in the form of a man,
and struck her several blows on the face and head :
by which she was bruised and torn in the way that
she was seen the next day by *Thomas Sohier.* And she
believed that the cause of this maltreatment was
because she would not go with the Devil to the house
of the said *Gallienne.*

She never went to the Sabbath except when her
husband remained all night fishing at sea.

Whenever she wanted to bewitch anyone and her
powder happened to have been all used up, the Devil
appeared to her and told her to go to snch a place,
which he named, for some more, and when she did
so, she never failed to find it there.

THE END.

www.ingramcontent.com/pod-product-compliance
Lightning Source LLC
Chambersburg PA
CBHW022032190326
41519CB00010B/1686